THE FORTH BRIDGE

Photography by Colin Baxter
Text by Jim Crumley

Colin Baxter Photography Ltd., Grantown-on-Spey, Scotland

A FAIRY TALE OF SCIENCE

It was the progeny of disaster. But it lives on, well into its second century now, deified by posterity as a monumental engineering wonder, beautified by floodlighting, and wondered at by succeeding generations of natives and tourists, proof if proof were needed of how well the lessons of that disaster were learned.

The collapse of the Tay Bridge in 1879 with the loss of 75 lives caused drastic reappraisals of a proposed Forth Bridge. At the time the bridging of the two great east coast firths was the obsessive ambition of the irresistible power of the railway age. The response when it came was a design by John Fowler and Benjamin Baker in 1881, a stark acknowledgement of engineering strength and quite without any gesture towards architectural aesthetics. Yet it took an architect, Alfred Waterhouse, to provide the first glowing endorsement of that very aspect of the bridge's character. In a letter to Fowler, he would write:

'One feature especially delights me – the absence of all ornament. Any architectural detail borrowed from any style would have been out of place in such a work. As it is, the bridge is a style unto itself; the simple directness of purpose with which it does its work is splendid and invests your vast monument with a kind of beauty of its own, differing though it does from all the beautiful things I have ever seen.'

And the verdict of posterity agrees.

We who have become accustomed to living with its imprint on the landscape of the Firth of Forth love it unquestioningly. It is massive, but it is also massively beautiful. It is a monument to brute strength but it is brutishly beautiful. It so

A monumental engineering wonder, beautified by floodlighting.

The bridge is a style unto itself... its simple direct- ness is splendid (opposite).

The Forth Bridge has become landscape itself.

demands to be contemplated and admired that it has become landscape itself, as organic an element (it seems) of the Forth as Berwick Law, the Bass Rock, or the dark profile of Edinburgh on its ancient stone plinth.

Yet because the Forth Bridge is so free of architectural contrivance, so pared to the bone in its design, it is architecture defined, or at least re-defined. It is a building, it has a unique form, it makes a unique imposition on its landscape. It IS architecture. It is more. It is a unique marriage of art and science and what more can you ask of architecture than that? It is a cathedral of engineering.

It is a building. It has unique form (opposite).

If you cared to define its shape at all, you might settle for diamond-shaped, three dull red and blunt-edged diamonds, hewn from a race of diamond giants. They wade the firth on colossal quadruped feet. You see, the temptation to breathe life into it, to wish a personality on it, is a strong one. Sheila Mackay, writer and publisher, has even feminised it in this passage:

'With the passing of the steam age and the arrival of the upstart road bridge, the Forth Bridge lost some of her glamour. Even Scots who threw coins out of carriage windows into the sea for luck countless times throughout childhood began to take her for granted.

'"The Bridge" seemed like an old-fashioned ornament on a cupboard shelf, still useful and too laden with memories to throw out, but somehow too incongruous to fit easily into the computer

4

age. She was built for steam trains. The sealed carriage windows of Intercity 125s discourage penny-throwing as we glide rather than chug across to Fife and back.'

And yet...

'For Scots everywhere "The Bridge" will always be an integral part of life's landscape, affectionately honoured and proudly celebrated as she lumbers into her second century.' (*The Forth Bridge*, Edinburgh 1990.)

She does lumber, too, in the heavy-footed way she wades the waters of the firth. And now that you have your eyes on the bridge's elephantine feet, you suddenly become aware of its third dimension. The thing has breadth. If you could now win the privilege of sailing beneath the bridge, or – better – step out on to the diamond-bisecting level of the railway track you would gasp at the girth of what soars above you. At track level you find yourself in an endless grove of steel trunks – redwoods – or a mile-long nave of leaning columns. You catch yourself walking softly, as you would in a cathedral, and looking high.

At such close quarters, little of that teeming skeletal architecture seems to make much sense. You are in the presence of a kind of rigidly ordered magic. You appreciate for the first time what designer Benjamin Baker had in mind when

he called it 'a romantic chapter from a fairy tale of science'.

Almost always, you look up to the bridge. Even if you walk out on the uncertain pavement of the road bridge, you reach no higher than waist level. It is the prerogative of the few to linger above it. Colin Baxter has done just that in his frequent explorations of the bridge's photographic possibilities. One of his images from almost directly overhead is surely among the most telling

'*... you would gasp at the girth of what soars above you.*'

A glimpse of the upstart road bridge between the bridge's elephantine feet (opposite).

that more than a hundred years of photographers have ever produced. The bridge seems to bury its own reflection deep into the Forth on still days, but the idea that it should throw a shadow is not one that would occur to many of us. And what a shadow! In a lowering western sun, the one thing which out-bridges the bridge is the shadow of the bridge for it distorts the height of the diamonds and accentuates their curved undersides and emboldens that most ruthless of engineering constructions with dark abstractions.

We cross it now without a thought. We know it isn't going to fall down. It has been a part of us for too long. Who would have thought that a hundred years after the most reluctant and nervous passengers began to use the bridge, we would be celebrating its centenary by, among other things, buying used rivets encased in glass as souvenir paperweights?

It has become a part of the language. An endless task is 'like painting the Forth Bridge'. It has starred in films. Its unmistakable form is world famous, as recognisable as the pyramids, if slightly younger.

There are signs, too, that a tide is turning against the motor car and towards public transport, with a crucial role for the train… Perhaps a new railway era is dawning, in which case the Forth Bridge stands as the great pre-eminence of old glories. No one will ever build anything like it again. Science appears to have run out of fairy tales.

HE WHO MAKES FEWEST MISTAKES

His name was Thomas Bouch, and if one man ever encapsulated the heady Victorian heyday of the railway era it was he. The quite unstoppable momentum which the industry had achieved by the 1870s elevated him onto a pedestal of endeavour and achievement, and by 1880 it had killed him.

His crowning glory was the Tay Bridge. It stood for just two years before a great storm bent it and broke it and splintered it. The train which was crossing the bridge at the time fell into the Tay and 75 people died. The following year – 1880 – Sir Thomas Bouch, as he had become, was scathingly belittled by a court of inquiry. His bridge was 'badly designed, badly constructed and badly maintained… For these defects… Sir Thomas Bouch is, in our opinion, mainly to blame.'

By then he had designed a suspension bridge for the Forth, and no single project of the great railway era was ever more emphatically doomed. It was all too much for Bouch, who died in November 1880.

But the railway companies were as ruthless as they were enterprising. The Tay Bridge was rebuilt and new design criteria for a Forth Bridge were laid down. Investors and potential passengers needed the reassurance of an immensely strong bridge. By June 1881, the Forth Bridge Company found what they were looking for, the design by John Fowler and Benjamin Baker which the whole world has since come to know and love.

Bouch had built hundreds of miles of railway in Scotland, he had designed floating platforms to carry trains across the great firths, but the towering scope and awesome strength of the Forth Bridge owes its character to the disaster which befell his Tay Bridge in December 1879.

Fowler and Baker were under no illusions about what was at stake. Baker wrote:

'If I were to pretend that the designing and building of the Forth Bridge was not a source of present and future anxiety to all concerned, no engineer of experience would believe me. Where no precedent exists, the successful engineer is he who makes the fewest mistakes.'

With the court of inquiry verdict on Bouch doubtless ringing in their ears, Fowler and Baker chose to build the biggest railway bridge in the world, not with iron but with the relatively new material of steel. And they gave an ancient principle of bridge building a new currency. Suddenly the word on everyone's lips was 'cantilever'.

In a comprehensive 'biography' of the bridge produced in 1890 by one of its engineers, Wilhelm Westhofen, the pedigree of the cantilever principle was spelled out. Far from being 'a modern and patentable invention… as a matter

The one thing which out-bridges the bridge is the shadow of the bridge (opposite).

of fact it is a pre-historic arrangement... In the oldest, as in the most modern wooden bridges will be seen practically the same thing... Skeleton bridges on a similar principle have for ages past been thrown by savages across rivers. Perhaps one of the most interesting structures of this kind ever built is a bridge in Tibet constructed about 220 years ago...'

If Fowler and Baker ever baulked at the enormity of the enterprise over the years of construction, they would have all the stimulus they needed on a rocky outcrop in the growing, darkening shadow of their embryonic bridge. For Bouch had also designed a Forth Bridge and it

was only abandoned after the Tay Bridge disaster and not before work on the first pier had begun. A scrap of it remains, symbolic of the price to be paid for getting it wrong.

Bouch's design had been for a suspension bridge, as had an earlier design of 1818 by James Anderson. Westhofen scoffed:

'To judge by the estimate the designer can hardly have intended to put more than from 2000 to 2500 tons of iron into the bridge, and this quantity distributed over the length would have given the structure a very light and slender appearance, so light indeed that on a dull day it would hardly be visible, and after a heavy gale no longer to be seen on a clear day either.'

The public, above all, wanted safety, reassurance, strength, stability. When the thing was done, when it was opened by the Prince of Wales on 4 March 1890, they had what they wanted. It wasn't just pronounced safe by its builder, it looked safe. But still the habit developed of throwing coins from the carriage windows into the firth, just in case. For they wanted not just a safe bridge, but also a lucky bridge.

AN ENGINEERING VOYAGE INTO THE UNKNOWN

The site for the crossing was perfect, the one lasting contribution to the project made by Bouch. Sheila Mackay summarised it thus:

'The natural features offered by the chosen site were excellent: above both shores of the firth, the land was sufficiently high to carry the railway lines on to the approach viaducts of the bridge while leaving adequate headroom for the largest naval and merchant ships; the water depth would allow the caissons, a vital element of the foundations, to be sunk; the rocky promontory of the Fife or north shore offered a bedrock for one of the cantilevers as well as anchorage for barges and launches connected with the construction work; the whinstone island of Inchgarvie overlaid with a bed of hard boulder clay (which Providence has so kindly placed in the middle of the firth, said Westhofen) acted like a gigantic stepping stone on which to construct the Inchgarvie cantilever. A more promising site was difficult to imagine...'

It took seven years to build. In the process, it killed 57 men and injured 461 others to a greater or lesser degree, an acceptable risk, it seems, from the point of view of the designer, the engineer, the company, and not least the men themselves

who daily confronted the knowledge and the danger of what they were building and where they were building it.

Whether they were perched 360 feet in the windy air on top of the cantilevers, or toiling in unthinkable claustrophobia deep beneath the water (and literally digging away the ground from under their feet), or anything in between those extremes of construction, danger and death shadowed their daily lives.

Benjamin Baker summarised:

'It is impossible to carry out a gigantic work without paying for it, not merely in money but in men's lives.'

No one argued.

It is difficult to look at the bridge today, now

The island of Inchgarvie acted like a gigantic stepping stone.

The Forth Bridge from the South shore, 9 August 1887 (opposite).

that it has long since settled into its landscape, even defined its landscape, and imagine the scale of the upheaval which descended on the shores of the firth. Queensferry, once no more than an inn and a street and the place where you could find a boat to take you across the sometimes loathsome waters of the Forth, became, in 1883, the place from which an engineering voyage into the unknown was launched.

The scale of the operation, the clamour, the arrival of a workforce which at times totalled more than 4000, the amassing of unthinkable quantities of materials, the buildings of houses and workshops – often huge workshops – (a hasty conversion was even inflicted on a medieval castle built by James IV on Inchgarvie, an untranquil billet for some of the workmen involved in the dangerous labours of the bridge's basement)... all that began to unfurl its massive chaos before the disbelieving eyes of the locals and growing numbers of curious onlookers and tourists.

They all knew there was to be a bridge, of course, but as the thing grew and grew hugely and over months and years, realisation of the bridge's monumental might began to take root. To see – and hear – it grow was to attend the birth of one of the great defining moments of engineering. Nothing less than that.

We tend to think of bridge builders as people with a head for heights, the balance of tightrope walkers and an indifference to colossal winds. But before such heroes can secure a single rivet the superstructure needs a base to anchor it to the land and the seabed. The key to the Forth Bridge's incomprehensibly strong foundations was the caisson, a wrought-iron cylinder, a 400-ton, 70-feet wide cylinder, that is. Because, of the Forth Bridge's twelve 'feet', six are under water.

The caissons were where the most hideous labours imaginable were carried out. They were floated out, filled with concrete which sunk them to the seabed, but it was the cutting edge of a projecting steel wall which rested on the bottom. Seawater was pumped out of the space between the bottom level of the concrete and the seabed, and replaced with compressed air. What was left was a space, seven feet high, and into this grim coliseum, men descended and began to dig away the very thing they were standing on until they bared the bedrock on which the brute mass of the bridge could rest securely.

Westhofen opined that such work demanded 'good health, freedom from pulmonary or gastric weakness, and abstemiousness, or, at any rate,

moderation in taking strong spiritous liquors. Some of the most experienced hands suffered when they had been making too free with the whisky overnight, and a good deal of the disorders that ensued were traceable to the same source; though, on the other hand, wet feet or incautious and sudden change from a heated atmosphere into a cold, biting east wind, insufficiency of clothing, and want of proper nourishment had their influence in causing illness among the workers.'

The influence of strong drink is a recurring

'The columns were perceived to be throwing out enormous far-reaching growths on either side.' Photograph taken 16 February 1888.

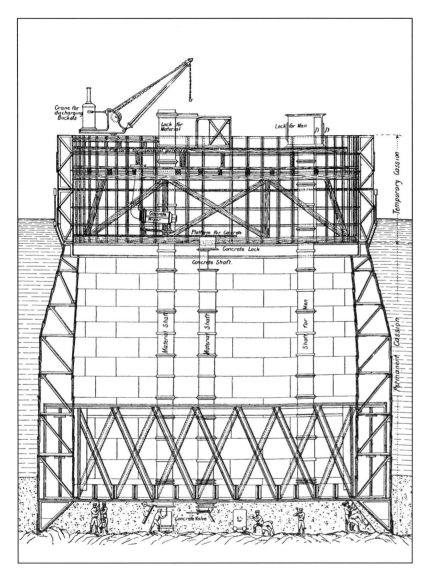

A section of one of the bridge's massive caissons, a vital element of the foundations, with its air-locks and claustrophobic working chamber (above). The brute strength of the bridge never fails to impress (opposite).

theme among the annals of the Forth Bridge construction. Baker noted:

'The Hawes Inn flourishes too well for being in the middle of our works, its attractions prove irresistible for a large proportion of our 3000 workmen. The accident ward adjoins the pretty garden with hawthorns, and many dead and injured men have been carried there, who would have escaped had it not been for the whisky of the Hawes Inn.'

Heaving the massive superstructure of the Forth Bridge into the sky was the greatest wonder of it all to the eyes of the ever growing crowds on the shores. It began with the columns for the Queensferry and Fife cantilevers. As these grew and grew, the public would see for the first time the height and therefore something of the true scale of what was involved. Each cantilever grew from four skewbacks, a fearsomely complicated joining together of five colossal steel tubes, and the means by which the entire weight of the superstructure was planted irrevocably into the bridge's foundations.

Then, with the cantilever columns in place, they began to sprout steel limbs at the most awkward-looking angles. In a biography of the constructor, William Arrol, Sir Robert Purvis would write:

'More and more was the amazement as week by week, the columns were perceived to be throwing out enormous far reaching growths on either side. Each of these was ever increasing in weight and altering in shape but ever in perfect balance.'

One of the most astonishing facets of this bridge of many astonishments was that because it was made in sections on the shore at Queensferry and all the rivet fixtures tested there before they were shipped out to the bridge site, the whole thing was actually built twice. Such was the obsessive attention to detail, such the preoccupation with strength and safety.

It was nothing less than characteristic of the whole endeavour, not to mention the era in which it was built, that the cantilevers should reach their full height in 1887, Queen Victoria's Jubilee year. Here was the ultimate piece of Victoriana, the era's grandest engineering gesture, and what could be more appropriate than that the grandest gesture should unfurl its own grandest gesture?

It became something of a fixation among the commentators of the day to grab an opportunity of ascending the cantilevers that Jubilee year to describe the view from the top. The best, or at least the most quoted, was inevitably Westhofen's, an account which confirms a deeper, poetic relationship with the bridge than was strictly necessary for an engineer:

'The broad river itself, with craft of all sorts and sizes, in steam or under sail, cutting across the current on the tack, or lazily drifting with the tide, is always a most impressive spectacle upon which one can gaze for hours with an admiring and untiring eye. And such it is, whether viewed in the glory of sunrise or sunset, in broad daylight with the cloud shadows flying over the

View from the north shore as the gap narrows, 24 May 1889 (opposite). The colossal profile of the Queensferry north cantilever from below and the Inchgarvie main pier, photographed 15 April 1889 (above).

surface, and a thousand ripples reflecting the sun's rays in every conceivable shade of colour, or in the soft haze of a moonlight night. The sunsets in summer are always magnificent, whether due to Krakatoan volcanic dust or to the vapours of the distant Atlantic, but there have also been many sunrises in early autumn when a hungry man could forget the hour of breakfast, and one could not find the heart to chide the worker who would lay down his tools to gaze into the bewildering masses of colour surrounding the rising light of day.'

As for Jubilee night itself, it was the bridge which provided the light and all the spectacle the celebrating crowds on the shore could wish for. Westhofen, inevitably:

'...the great masses of the central towers of the bridge lighted up by hundreds of electric arc lights – Lucigen and other lamps – at various heights where work was carried on, formed with their long-drawn reflections in the waters of the firth, three pillars of fire, and afforded a truly wonderful and unique spectacle.'

So the three great cantilever shapes grew and solidified and linked arms and the gap was bridged. The Prince of Wales performed the opening ceremony on 4 March 1890, in a gale which precluded the possibility of any speech on the bridge itself. He screwed in the last rivet with a silver key, and with the words, 'Ladies and gentlemen, I now declare the Forth Bridge open', the thing was done.

A STRUCTURE
SO STUPENDOUS

The Forth Bridge statistics are almost as famous as the bridge itself. The Prince of Wales spoke at a post-opening ceremony lunch:

'It may interest you if I mention a few figures in connection with the construction of the bridge. Its extreme length including the approach viaduct, is 2765 yards, one-and-one-fifth of a mile, and the actual length of the cantilever portion of the bridge is one mile and 20 yards. The weight of steel in it amounts to 51,000 tons and the extreme height of the steel structure above mean water level is over 370 feet, above the bottom of the deepest foundation 452 feet, while the rail level above high water is 156 feet.

'About eight millions of rivets have been used in the bridge, and 42 miles of bent plates used in the tubes, about the distance between Edinburgh and Glasgow.

'Two million pounds have been spent on the site in building the foundations and piers; in the erection of the superstructure; on labour in the preparation of steel, granite, masonry, timber and concrete; on tools, cranes, drills and other machines required as plant; while about two-and-a-half millions has been the entire cost of the structure, of which £800,000 has been expended on plant and general charges.

'These figures will give you some idea of the magnitude of the work, and will assist you to realise the labour and anxiety which all those connected with it must have undergone. The works were commenced in April 1883, and it is highly to the credit of everyone engaged in the operation that a structure so stupendous and so exceptional in its character should have been completed within seven years.'

And then of course there is the matter of the paint: 35,527 gallons of paint oils and 250 tons of paint were required while the bridge was being built. It was a good career move for the Leith firm of Craig and Rose when they won the contract to supply the paint, for they supplied all of it for the next one hundred years. The very shade of muted red they used became known throughout the world of paint as 'Forth Bridge Red'. But privatisation of the railways changed all that. Railtrack, the company which superseded Scotrail as the owner of the track (and therefore the bridge) but not the trains which run on it, came into being at a time of anxious and persistent criticism of the condition of the bridge and the standard of maintenance. The longest memories could never remember it looking so run down and rusty. The longest memories, aided by eager historians, pointed to the folly of cutting corners and costs where bridge safety was concerned.

No one was actually suggesting that the Forth Bridge was going to fall down, but matters were brought to a head when the West Lothian MP

Tam Dalyell took the issue to the floor of the House of Commons. The upshot was a reappraisal of the paint contract, a new supplier, and, befitting the new era into which both bridge and railways were moving, a change of paint. Out went the old Craig and Rose formula, and in came one already in use on North Sea oil rigs, where fending off the viciously corrosive effects of open sea and salt winds is the first priority.

As for the idea of painting the bridge from end to end, a task so time-consuming that by the time you have finished you need to start again at the other end, it just doesn't happen that way, and apart from the very first painting, it never has. But if it did, the bridge's 145 acres of painted surfaces might need 7000 gallons. What does happen is that the most exposed areas are treated more often than the rest, but one way or another it all gets painted about every five years.

Some of the statistics are so odd that you wonder who bothered to count. The rescue boats stationed under each cantilever, for example, saved eight lives and recovered 8000 caps which had been blown from workmen's heads into the water. No one has recorded what happened to the caps once they were recovered. There are still rescue boats for bridge workers, even though the workmen's caps have been replaced by hard hats. And some of the workmen are now women.

But the ultimate statistic is the one the Forth Bridge insists upon by its very survival. That it is simply one of a kind.

THE WONDER OF ITS OWN FUTURE

Time has imparted timelessness. You look at the Forth Bridge today, floodlit and famous and still perfectly functional, and you cannot imagine the river without it.

When it was built, it bridged not just the shores of the Forth but eras of bridge building. Baker was a daring, forward-thinking designer working at the cutting edge of his profession. Arrol's craft was one of long tradition. What Sibelius did for the symphony and Cézanne for painting, the Forth Bridge did for engineering and industrial architecture, drawing on the past and wedding tradition to daring to create something unprecedented.

It lives on, a working survivor of its own history. It survived bombing raids during World War II. It survived the advent of the road bridge in 1960-something-or-other... you see? Who remembers the age of a road bridge? It has even survived the prospect of a second road bridge upstream from the first, now that the first one is not quite up to the demands being put upon it.

Yet as our gaze begins to wander away from our obsessive relationship with the motor car and to contemplate alternative possibilities to moving ourselves and our industrial goods from one part of the country to another, our eye falls again with something between pride and affection on the great familiarity of THE Forth Bridge. It has never failed to handle the demands put upon it, and as long as we are prepared to maintain it scrupulously, it never will. Railways are back in our train of thought again after years of decline, and for as long as we need trains we will need bridges.

The Forth Bridge is the supreme ambassador for railway bridges everywhere because it is more than a bridge. It has become heritage in its own working lifetime. It has even become the subject of its own permanent exhibition in the Queensferry Lodge Hotel near the north end of the Forth Road Bridge. It is no coincidence that although the exhibition professes to honour both bridges, the hotel itself chooses a logo based on the railway bridge with a steam locomotive in full flight. Even on

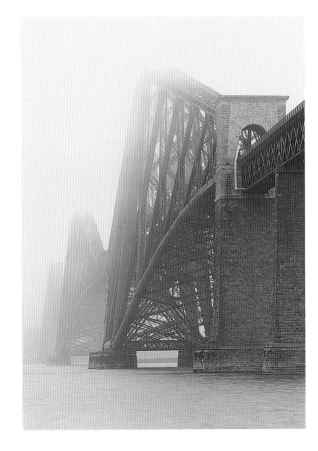

A sea mist and high viewpoint ...engineering revealed as art. The supreme ambassador for bridges everywhere.

the frontiers of twenty-first-century tourism, the great icons of the past are an irresistible lure.

Society has finally agreed, it seems, that the bridge is both engineering and architecture. It is honoured by the Institution of Civil Engineers and the American Society of Civil Engineers as an Historic Engineering Monument. And, in the system used for protecting and categorising the merits of works of architecture, it is officially recognised as Scotland's biggest 'listed building'. So after a debate lasting many decades, the issue is settled – it is the best of both worlds.

The one thing the Forth Bridge has never become is a museum piece, and it need never become one. It was the wonder of its day when it was built; now it is the wonder of its era. But as its era is far from over, it is also a wonder of its own future. It has a way of repeating its own history. For example, the restoration work on the towers begun in 1998 was serviced from Longcraig Pier at Queensferry, the very pier which was rebuilt – to today's proportions – to service the construction of the bridge, so that Queensferry's ferry traffic and other firth-plying business could carry on undisturbed by the bridge work.

The Forth Bridge may have been the progeny of disaster, but it was also the sublime realisation of an ancient vision. It remains above all its own monument to the visionary daring which designed it and (as with all cathedrals) to the anonymous thousands who laboured to make it what it has become.

FURTHER INFORMATION

Viewing Points:
North Queensferry Tourist Information Centre,
North Queensferry Lodge Hotel, St Margaret's Head,
North Queensferry, Fife KY11 1HP. Tel: 01383 417759.

There is an excellent viewing point here for both the Forth
Rail and Road Bridges, as well as a permanent exhibition
situated in the hotel. The esplande in South Queensferry
also provides close-up views for both bridges.

Excursions:
Regular boat trips from Hawes Pier on the esplanade in
South Queensferry sail under the Forth Bridge to Inchcolm
Island in season. For further information, contact
Maid of the Forth, Tel: 0131 331 4857.

Other Sights:
Queensferry Museum, 53 High Street, South Queensferry.
Tel: 0131 331 5545. Explanatory displays.

The Hawes Inn, at the east end of the esplanade in South
Queensferry, dates from the 17th century. It was a popular
public house in the days of the bridge's construction and
features in the novels of Robert Louis Stevenson.

First published in Great Britain in 1999 by
Colin Baxter Photography Ltd
Grantown-on-Spey, Moray PH26 3NA, Scotland

Photographs © Colin Baxter 1999
Text © Colin Baxter Photography Ltd. 1999
Reprinted 2003
All rights reserved.

The archive photographs on pages 10, 12, 15, 16, 18, 19 are reproduced from original prints held
in the History Collection of the Civil Engineering Library at Imperial College, London. They are
reproduced with the kind permission of Mrs K Crooks, Departmental Librarian.

A CIP Catalogue record for this book is available from the British Library.

ISBN 1-84107-001-7   Printed in Hong Kong

The Forth Bridge remains its own monument
to the visionary daring which designed it.